621

PICTURE FACTS

EARTHMOVERS

D0784028

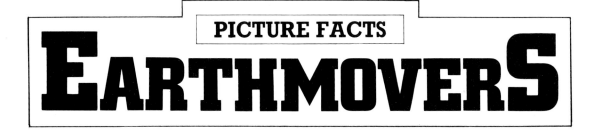

R. J. Stephen

Franklin Watts

London New York Sydney Toronto

Published by:

Franklin Watts
96 Leonard Street
London EC2A 4RH

Franklin Watts Australia
14 Mars Road
Lane Cove
NSW 2066

ISBN: Paperback edition 0 7496 0507 3
Hardback edition   0 86313 400 9

Copyright © 1986 Franklin Watts

Paperback edition 1991

Hardback edition published
in the Picture Library series.

Printed in Singapore

Designed by
Barrett & Willard

Photographs by
Atlas Copco
Bucyrus-Erie
Case Poclain Marketing
Caterpillar Overseas
CPE/Norman Webber
Geoff Mead
Intercontinental Dredging Company
JCB
Krupp Rheinhausen
NEI Thompson
Roughneck Excavators
State Electricity Commission
   of Victoria

Illustration by
Rhoda and Robert Burns

Technical Consultant
Geoff Mead

Series Editor
N. S. Barrett

Contents

Introduction

An earthmover is a machine with a blade, bowl or bucket. It is used for excavating, or digging up material such as soil, rock or dirt.

There are many types and sizes of earthmovers. They carry out all kinds of building work, big or small. Some of them are called excavators.

△ An excavator at work on a building site. The operator can turn the upper half of the machine in a complete circle. The shovel scoops earth back towards him.

Earth has to be moved whenever building work is being done. Houses and other buildings need foundations, which must be sunk into the ground. Roads are built through hills and fields. Tunnels are bored and the bottoms of rivers are dredged. All kinds of materials are mined and quarried.

△ A power shovel unloading material onto a dump truck. The shovels, or buckets, on these machines are big enough to drive a car into.

The excavator

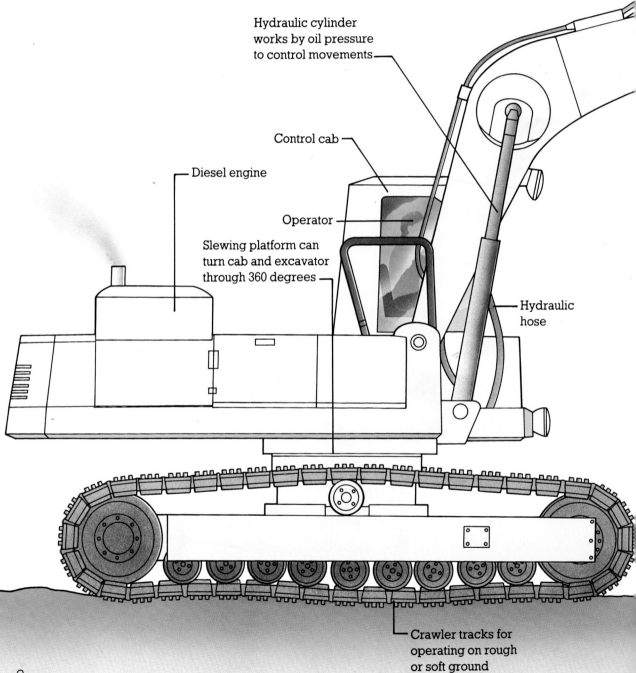

Hydraulic cylinder works by oil pressure to control movements

Control cab

Diesel engine

Operator

Slewing platform can turn cab and excavator through 360 degrees

Hydraulic hose

Crawler tracks for operating on rough or soft ground

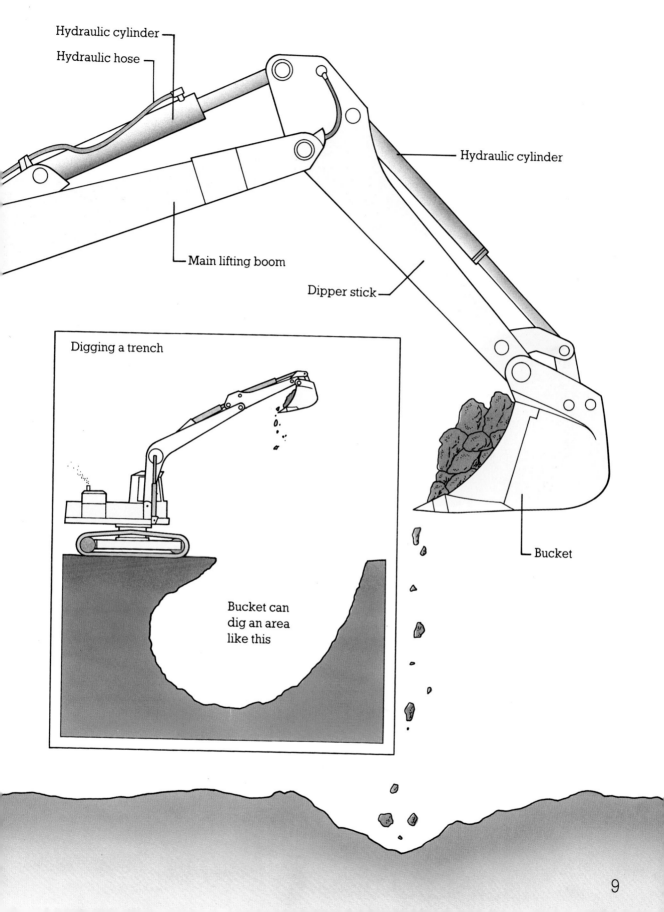

Hydraulic cylinder

Hydraulic hose

Hydraulic cylinder

Main lifting boom

Dipper stick

Digging a trench

Bucket can
dig an area
like this

Bucket

Working an earthmover

The person who works, or operates, an earthmoving machine sits in a cab similar to that of a truck. But there are more controls to work. The operator's cab of a giant power shovel might be situated 20 metres (65 ft) above the ground.

Backhoe loaders have an attachment at the back as well as the front. The operator has a set of controls for each.

▽ Inside the cab of a walking dragline excavator. The operator has controls for smooth, accurate control of the bucket and for operating the walking mechanism of this giant machine. There are also switches for working the wiper, defrosters and floodlights, as well as safety warning lights.

An operator working a
backhoe loader. The
two legs, which are
operated from the cab,
hold the machine
steady while digging.
The operator can swivel
round to drive the
tractor and work the
loader controls.

△ The operator is just
starting to dig.

▷ The bucket scoops
out the earth.

Kinds of earthmovers

Earthmoving machines are mostly self-powered or are tractors with attachments. Bulldozers are crawler tractors with a heavy steel blade in front. Loaders have a bucket for transferring the material to dump trucks or other vehicles.

Backhoes and power shovels have a bucket attached by a long arm called a boom. A dragline excavator pulls a bucket by a cable attached to a winch.

△ A track-type crawler loader (left), a 360° excavator (behind) and a wheel loader (right) emptying material into a dump truck.

◁ An excavator used for clearing waterways such as drainage systems. It has a long boom for stretching across to rake in vegetation and floating debris. The boom is pulled in by a cable.

▽ Bulldozers working on a dam. They have a heavy steel blade which clears away boulders.

▷ Most backhoe loaders working on large building sites can be used for several purposes. Here, the machine is being operated as a loader. At the rear is a backhoe, which can dig deep trenches.

Backhoe loaders such as these operate by hydraulics, using oil pressure. More than a tonne of material can be lifted.

Many kinds of earthmovers work on sites being prepared for roads, housing projects or other building construction. Most of them have a boom with two or three hinged joints. The boom carries the digging implement.

Loaders fill dump trucks with the excavated material. Dump trucks are sturdily built, with large knobbly tyres for working on rough ground. They unload the material by tipping.

△ Motor graders grade and shape the surface of the ground. They have a long blade mounted underneath. The operator can adjust the blade to work in various positions.

▷ An excavator drops its load into a dump truck. This one runs on crawler tracks so that it can operate on soft, rough or rocky ground.

▽ The cab and boom of this excavator are mounted on a slewing platform. This enables the operator to turn a complete circle above the tracks. He is operating a backhoe on the face of a slope.

◁ A walking dragline excavator looks very much like a crane. It has a digging bucket hanging from a cable at the end of the arm, or jib. The dragline is the cable that runs from the base of the jib to the bucket. It can be winched in, dragging the bucket through the earth.

Dragline excavators usually move on tracks. They are used for digging canals, ditches or other large holes. They work at the top of the excavation and drag their bucket upwards.

△ A dump truck is a vehicle used to take away the excavated material. These trucks are very strong, with large, heavy wheels. Some dumper trucks can carry 100 tonnes or more. Small dumpers used on building sites might carry less than a tonne.

◁ A skid-steer loader. Its wheels on either side can be braked so that it skids round. This makes it extremely manoeuvrable.

△ A bulldozer with a ripper at the back. The sharp tooth is pulled through the hard ground to break it up.

▷ A special attachment on the backhoe of this machine is used for boring holes.

Bucket-wheel excavators

Bucket-wheel excavators are designed to cut away material from the sides of a slope. They operate continuously and are used for such purposes as open-cast coal-mining.

The buckets are mounted on a wheel at the end of a long, movable boom. As the wheel is rotated, the buckets fill up and then deposit their load onto a conveyor system that carries it away.

▷ A close-up look at the wheel and buckets of a giant excavator. This massive machine moves on tracks.

▽ A bucket-wheel excavator at work on an open-cast coal-mine. It is used for removing the layers of earth above the coal as well as for digging out the coal. It stands as high as a 30-storey building.

▷ Bucket-wheel excavators like this one are capable of mining 30,000 tonnes of coal every day. They weigh as much as 13,000 tonnes.

The wheel revolves continuously. As each bucket scoops up a load of material, another bucket on the other side of the wheel is emptying its load on to a conveyor system. This takes it across to a waiting railway truck. At some mines, the coal is conveyed directly to a power station.

Dredgers

Dredgers are floating excavators. They clear the mud and silt from the bottom of rivers, harbours, canals and other waterways. Some dredgers are sea-going ships.

Some dredgers lower buckets or grabs to the bottom of the water to scoop up the dirt and mud. Dipper dredgers have a large scoop at the end of a chain. Suction dredgers work by sucking up loose material.

△ A bucket dredger in action. A chain of buckets (centre) is raised and lowered on a frame called a ladder.

This type of dredger usually has to be towed to the site by a tugboat. It discharges the dredged material into barges – one can be seen in front. The material is then hauled out to sea and dumped.

Building tunnels

Special machines are used for boring tunnels. Tunnels through solid rock can be blasted out with explosives, because no support is needed for the top and sides.

Machines called moles are used for digging tunnels through soft earth or clay, or under a river bed. The face of the mole is a revolving disc with cutters. The excavated material is carried away by a conveyor belt.

▽ The cutting tool at the front of this tunnelling machine bores through the earth. The waste material passes back to a conveyor belt to be carried away.

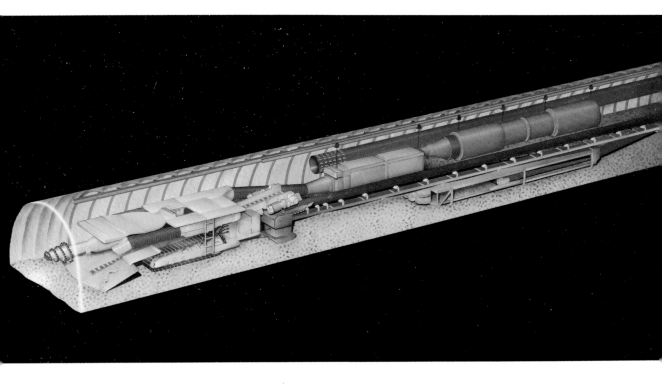

The story of earthmovers

Manpower

For thousands of years, all the earthmoving needed for building houses or making roads was done by people. Thousands of slaves were used to build the pyramids of ancient Egypt. Their tools included picks and spades, and wheelbarrows and similar containers. Animals, such as horses and donkeys, helped to transport material or perhaps to work a treadmill.

△ A revolving steam shovel of around 1915, before the age of dump trucks.

The first machines

The first earthmoving machines were built for dredging rivers and digging canals, in the 1500s. Even when steam power first became available, in the late 1700s, it was used to work dredging machinery. Human labour was plentiful and cheap, and was used for earthmoving tasks for many more years.

△ A coal loader of the early 1920s. It loaded by thrusting forward. Behind it is a giant stripping shovel.

Digging across the States

When the United States expanded westwards in the 1800s, large-scale earthmoving projects were undertaken, such as the construction of railways. This required land-based excavators. William Otis of Philadelphia designed the first of these machines in the 1830s. It

△ A walking dragline working on the Welland Ship Canal, a major Canadian waterway which was completed in 1932.

was a steam-powered shovel, which could break and remove material with a single bucket.

Mechanization

The digging implements used by workers or drawn by horses gradually became fully mechanized. Steam was used to power such earthmovers as the scraper, which loads, transports and then dumps.

Crawler tractors were also developed and used as the basis for bulldozers and loading shovels. Wheeled loaders and bulldozers were introduced in the 1920s, backhoe loaders in the 1940s.

The first walking dragline, on the other hand, came into use as early as 1913 and diesel engines were first used in 1923.

Hydraulics

The early earthmoving machines had cables for moving the various parts. Hydraulics, which uses the pressure of oil, was first developed in the 1940s. Small excavators are now all operated by hydraulic systems. Large hydraulic excavators and mining shovels have also been built, but cables are still used to operate some of these monster machines.

△ A modern crawler excavator is operated by means of hydraulics. The cylinders which work by oil pressure, move the working parts of the excavator.

Mini excavators

Earthmovers have become smaller as well as bigger. There are mini excavators for doing small jobs or working in limited space. One of the latest, known as a micro excavator, can be used for garden work and is small enough to get through a standard doorway. For use with it is a "micro dumper". Both may be towed behind a car.

△ A micro excavator (left) empties its load into a micro dumper.

Facts and records

Big Muskie

The world's biggest walking dragline is the Bucyrus-Erie 4250W, known as Big Muskie. It operates in the Ohio Power Company's Muskingham mine.

Big Muskie is the largest moving land machine in the world. When the boom is pointing forwards, the machine is 148.6 m (487.5 ft) long, $1\frac{1}{2}$ times the length of an average football pitch. With the boom pointing upwards, it stands 67.8 m (222.5 ft) high.

Big Muskie's bucket is larger than a two-storey house. It is the biggest bucket ever made. It digs 295 tonnes at a time. Three bucketsful would be enough to fill a 54-truck coal train.

The dragline uses more electric power than a city of 100,000 people.

△ The largest bucket-wheel excavator.

Largest excavator

A series of bucket-wheel excavators made in West Germany are 210 m (689 ft) long, 82 m (269 ft) tall, with a wheel measuring nearly 68 m (222 ft) in circumference.

▽ Big Muskie dwarfs all other machines.

Glossary

Backhoe
An earthmover that pulls its bucket through the earth.

Backhoe loader
A digging machine used chiefly on building sites. It has a backhoe at one end and a loader at the other.

Boom
The main arm of a digger.

Bulldozer
A crawler tractor with a heavy steel blade mounted in front.

Crawler
A machine that moves on tracks.

Dragline excavator
A machine that digs material by pulling a bucket with a wire rope.

Dredger
A machine for digging under water.

Dump truck
A vehicle that takes away material and dumps it by tipping.

Excavator
A digging machine used for bulk earthmoving. Smaller digging machines such as backhoe loaders are also often called excavators.

Hydraulics
A system for operating the digging tools of earthmoving machines. Hydraulic cylinder rods extend and withdraw with the force of oil pressure.

Loader
A loading shovel.

Power shovel
A mechanical bucket used for digging and loading.

Ripper
A large steel tooth, or set of two or three teeth, mounted on the back of a crawler tractor or bulldozer, used to break up rough ground.

Scraper
A vehicle used for bulk shifting of earth at high speeds and over long distances. It digs, loads itself and transports the material.

Slewing platform
On some earthmovers, this enables the operator to swivel the cab and boom round.

Index